S0-AUI-091

Simple enamelling

Fig. 1 *Wapiti* enamelled on steel tiles by Geoffrey Franklin. Bowes Museum, England (see page 37)

Simple enamelling

Geoffrey Franklin

2 4 2 6 7

PRESCOTT HIGH SCHOOL LIBRARY
PRESCOTT, ARIZONA

WATSON-GUPTILL PUBLICATIONS / NEW YORK

Acknowledgements

The author gratefully acknowledges the co-operation of the following artists and institutions in the preparation of this book: Henry Haigh (enamellist); Madge Martin (enamellist); Edward Winter (enamellist); Karl Drerup (enamellist); Kenneth Bates (enamellist); Joe Thomason (photographer); the British Museum; the King's Lynn Corporation; the Schools Museum Service, Bowes Museum; the Somerset College of Art; the Victoria and Albert Museum; the Worshipful Company of Goldsmiths; and the Blue School, Wells, Somerset.

Copyright © 1971 by Geoffrey Franklin
Published 1971 in London by Studio-Vista Limited
Published 1971 in New York by Watson-Guptill Publications,
a division of Billboard Publications, Inc.,
165 West 46 Street, New York, N.Y.

All rights reserved. No part of this publication
may be reproduced or used in any form or by any means — graphic,
electronic, or mechanical, including photocopying, recording,
taping, or information storage and retrieval systems — without
written permission of the publisher.

Manufactured in U.S.A.

ISBN 0-8230-4833-0

Library of Congress Catalog Card Number: 77-133981

First Printing, 1971

General Editors Brenda Herbert and Janey O'Riordan
Set in 9 point Univers
Manufactured by Parish Press, Inc.

Contents

Introduction

With the increase of leisure in modern times it is not surprising that the arts and crafts have become more widely practised than at any previous time. Painting, pottery and many others are becoming more widespread as creative activities which enable people to realize the latent talents within themselves. Enamelling has taken its place in this pattern and has never before enjoyed such popularity as it does today. Being one of the oldest crafts known to man, its present-day appeal is difficult to analyse. Certainly a major factor must be the development of industry which makes kilns and materials available within the purchasing range of most people. Appreciation of enamel has not diminished over the centuries, yet neither have its fundamental properties of fire, brilliance and subtlety of colour greatly increased. What has happened is that, with the development of communications, the trade secrets previously guarded by workshops and craftsmen and handed down from father to son, have now become available for anyone who wishes to exploit them.

Having taken the trouble to obtain this book, you have already shown an interest in enamelling. The next step should be to obtain materials in order to practise what has been learned. It is then that a real involvement with enamelling begins. I have heard it said that enamelling is infective, so absorbing is its fascination. Once you begin it is not long before you will realise that the masters, although having given up their many secrets, have left many as yet undiscovered. This book attempts to give an introduction to the enamelling processes, but the craft remains an individual one, with each enamellist creating not only his own work but also his own methods of working. The practice of enamelling may be taken as far as you wish. Immense satisfaction and real pleasure may be obtained from the simplest pieces which may be made quite quickly. Alternatively, more ambitious pieces may be undertaken which will consume more interest and time, yet still yield appropriate satisfaction and pleasure. When finished, an enamelled piece, given reasonable treatment, may remain perfect for over a thousand years, although it is the product of a few hours' or even a few minutes' work.

Anyone may enamel successfully; even beginners will achieve a worthwhile result for their efforts. The basic techniques require no greater skills than the average person can quickly acquire. It would be foolish to suggest that all the processes are elemental and that none require a high degree of skill. Advanced enamelling can be the most exacting of the art crafts, requiring the touch of a master to realize the full range and potential of the medium. However, many people have achieved results from enamelling which their previous craft aptitudes had not suggested were possible.

As a beginner, you can start enamelling for a small financial outlay. Alternatively, if you are fortunate enough to live in an area where classes in enamelling are available, you may attend them and thus avoid the initial outlay. If no such classes exist, it is as well to remember that, provided there is sufficient demand, schools and colleges will make every effort to provide these for you. Whichever way you choose, I hope that this book will be of some assistance in your enjoyment of this fine art craft and its beautiful range of colour.

1 The nature of enamel

Enamel is fundamentally coloured glass. The enamel is laid on a clean metal surface and is melted until it fuses with the metal at a temperature of approximately 750° centigrade, 1382° fahrenheit. This firing of the enamel is usually done in a kiln, but there is another method of firing which is described later in this book (page 64).

Enamel is commercially available, already ground in the form of a fine powder. In previous centuries craftsmen made their own enamels. Fortunately this is no longer necessary. To make and prepare your own enamels is a long, tedious and wasteful process in terms of time and material. The ready-made product is excellent in quality and in technical stability. Moreover the oxides and other materials required to make enamel are not always readily available. Unless you are an experienced enamellist requiring a particular hue or shade for a specific purpose, the cost and time involved would be prohibitive.

Enamel is divided into two distinct categories, opaque and transparent. Opaque enamels are, as their name suggests, nontransparent. Their fired finish is glossy, but this can be varied considerably by careful underfiring. This means allowing the enamel just sufficient heat to fuse it to the surface of the metal but insufficient to glaze the enamel. By this method a rich, coarse sand-like texture may be obtained. Other grades of this texture can be attained before the ultimate fire-glazed finish (see page 50). Opaque enamels are intermixable, thus giving the effect of another colour. By mixing a yellow with a blue, a very real effect of green is obtained. Just as the French impressionist painters of the nineteenth century painted spots of blue and yellow closely together to give the visual effect of green, so the same thing occurs with mixed opaque enamels, except that the dots are micro-grains of enamel and the effect is perhaps more convincing. This mixing of enamel should not be confused with the very wide range of manufactured colours available. It is not unusual for the beginner to confine his use of opaque enamel to these manufactured colours rather than mixing his own. Almost all colours are available and within these there is a wide selection of hues and tones.

Transparent enamels are coloured enamels which allow light to pass through them and to reflect and show the metal surfaces beneath. The basis of all transparent enamels is clear flux. This is a clear colourless enamel to which metal oxides are added in order to obtain the coloured but transparent appearance. Transparent enamels have a glossy fire-glazed finish. It is not possible to underfire them since until they are glazed their appearance remains opaque (see also pages 52 and 98–9). Nor is the inter-mixing of the colours to be encouraged. This may be achieved with practice with very few of the colours, but generally a cloudy, dull effect is the result, often with some separation taking place. Transparent colours are more delicate and subtle than opaques and enable you to work the metal surface before enamelling it. This can be done by a variety of methods; the most popular ones are engraving and the laying of fine metal foil underneath the enamel (see page 71).

The particular beauties of transparent enamels defy description. When used in conjunction with the opaques, and by slightly overlapping them, the enamels assume another tone and dimension of colour which the reader must practise and see to fully appreciate. It is the transparents which give the art of enamelling its true distinction.

The metal

Many metals may be enamelled upon. Gold and silver are the very best since they allow the most reflective and refractive surfaces for the use of the transparents. They also allow the enamel to fuse well and, because they are the so called 'noble metals', they work and form readily prior to enamelling. Naturally one cannot afford to use these metals every time but it is worthwhile considering their use when you have a little more experience, since you will then have something which is not only very beautiful but which is of some value.

After these, copper and gilding metal are the most useful and popular metals as they are not expensive. Copper is very suitable as it readily conducts heat and fuses well with the enamel. It offers a bright surface for the use of transparents, and cuts and forms easily before enamelling. Its disadvantage is that after firing it is in an annealed state, that is, in its softest condition, and rough usage of an enamelled copper article can distort its shape and cause the enamel to shatter. Its principle uses are for flat pieces and jewelry.

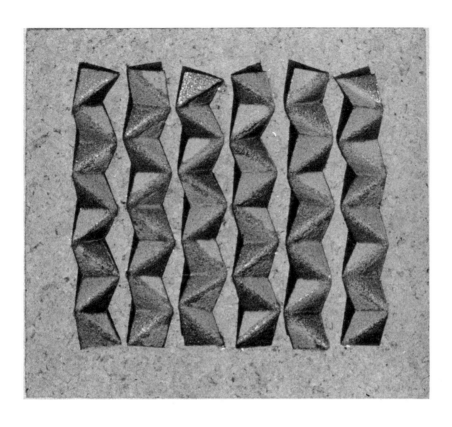

Fig. 2 Pre-formed pieces of sheet steel, enamelled and then cemented onto a cork tile. 1st-year student, Somerset College of Art, England

Gilding metal is popular since it offers a very bright surface for the use of transparents. It is a harder metal than copper and therefore a little more difficult to cut and preform, but this extra hardness makes it well suited for domestic items. After polishing, gilding metal can be mistaken for gold by the inexperienced eye. It has a quality all of its own and lends itself well to designs which do not cover the entire surface of the metal with enamel. Its disadvantage is that it is an alloy of copper and brass and as brass contains zinc, firings should be limited to two. Any more firings will tend to burn out the zinc content and will cause some discolouration, spoiling the finished piece. A wide range of preformed articles of both copper and gilding metal is available from enamel suppliers. These vary from ash trays to bangles and, being made specially for enamelling upon, are clean, free from solder

Figs 3 and 4   Enamelled steel panels by 1st-year students at Somerset College of Art, England (see also figs 21 and 22). The enamels were applied by the dusting and stencil methods described in chapter 3, with the occasional use of a spatula

and of a suitable gauge of metal. For someone who is not interested in forming their enamelled pieces these products are particularly useful (see also page 79).

Aluminium; enamels for this metal are now available in the United States. However, because of the low melting point of aluminium, the enamels have had to be adapted to fuse at this lower temperature. At the time of writing only a limited number of pastel colours are available. Given the correct design to utilize these pastel effects, enamelling upon aluminium is a creative possibility. The development of these enamels is primarily for industrial purposes. It is possible that in the future, with further development, a greater range will be available for the individual enamellist.

Steel; perhaps the biggest single advantage of steel is its low

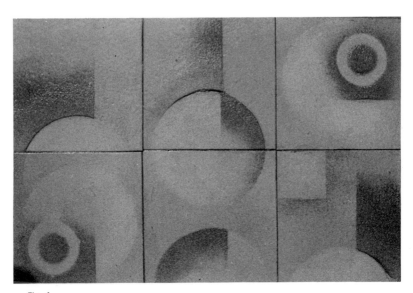

Fig. 4

cost. It is ideal for the use of opaques and the subsequent over-laying of transparents, but transparent enamels laid directly onto steel are not effective. It is a strong metal, excellent for flat pieces of work, but is difficult to join and preform. Ordinary mild steel can be enamelled upon provided the metal is thoroughly clean and the number of firings is restricted. A special enamelling steel is available and this should be used wherever possible. I have enamelled extensively on steel and use this metal with students in the early stages. It is not as prone to heat distortion as other softer metals and is therefore ideal for the beginner. When cut into tiles it is excellent for wall panels, murals and other flat items (see chapter 9).

With experience you will soon learn which metal to use with which enamels. Enamelling is such an individual art that it is impossible to be categorical on any single point. These notes are guide lines. It is up to the individual to to develop his or her own lines of work, remembering always that enamel is glass and that the metal parts are generally a means of presenting the enamels.

2 Basic equipment for the beginner

Before beginning to describe the various techniques of enamelling it is necessary to indicate the tools you will need in order to practise this art craft. These tools are all, with the exception of the kiln, inexpensive. Even the kiln need not be purchased immediately as quite good results can be obtained by using a gas torch (see page 64). However, if you are to progress beyond very small pieces of work, the purchase of a kiln will be essential.

Kilns

For the beginner I would recommend one of the medium size kilns in the small range. The other very small kilns are efficient but have a limited life. Parts such as electric elements and kiln walls cannot be replaced owing to the construction of the kiln. They have limited sized firing chambers and are, in effect, disposable kilns to be used mainly for test firings and jewelry.

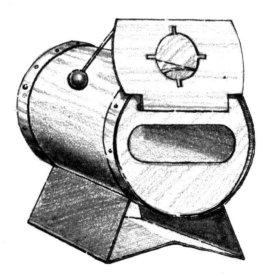

Fig. 5   A small kiln with a firing chamber of approximately 4 in. × 4 in. × 2 in. On this type the door opens upwards

The intermediate sized kilns are efficient and have the advantages of a few refinements. They allow the replacement of parts such as elements and, because of the larger size of the firing chambers, a larger and wider variety of work can be fired in them. The smaller kilns are inexpensive for the work they do but in practice the slightly larger ones at a reasonable price are the better kilns.

There are two main types of kiln, gas and electric. For the purposes of this book I will confine my comments to electric kilns, as these are more readily available in a wide range of sizes at competitive prices and are very easily installed. Gas kilns are comparatively rare nowadays and as a general rule are large and expensive for the purposes of the amateur. The ideal medium size electric kiln is one with a firing chamber of 10 ins × 10 ins × 5 ins.

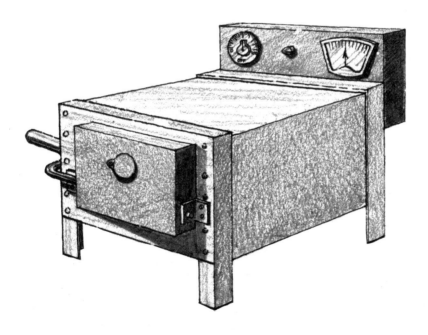

Fig. 6   A medium size kiln with a firing chamber of approximately 10 in. × 10 in. × 5 in. On this one the door opens downwards

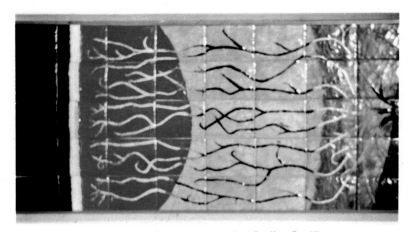

Fig. 7   Abstract composition on copper tiles. Geoffrey Franklin

Fig. 8   Part of an exterior panel by Henry Haigh, England

Depending on how much you can afford to pay, this is the size I would recommend. There are many other sizes available, all of which only require an asbestos pad to stand them on and to be plugged into your own domestic power socket to install and operate them. An electric kiln is economical to run. The smaller kilns run at approximately the same cost per hour as a one-bar electric fire (or a single burner hot plate), and the medium size ones at a little less than twice the cost. Whichever you choose, the following points should be considered before puchasing:

Stability of the kiln (a good firm base is essential).

A good size of firing chamber, allowing a reasonable range of work to be fired.

A well-insulated kiln case and power lead, both for electrical safety and to conserve the heat.

An adequate and reliable peep hole through which to observe the enamel firing. (Mica windows tend to become opaque after a few firings.)

A door which stays open and is well clear of the hand when putting work in the kiln, and with a well-insulated door handle.

There are many kilns which satisfy these basic requirements and allow easy replacement of worn out elements. Given reasonable treatment, a kiln of this type will last for many years.

Enamels

To begin with enamels should be purchased in ready ground form and ideally in 4 oz. containers. This saves a great deal of tiresome searching for suitable bottles and is the most economical quantity for the amateur to purchase. I would recommend the following colours to start with. These provide a good basic palette if you remember that the opaques are intermixable to some extent.

Opaques – Two varieties of red; two varieties of blue; two varieties of yellow; one black, one white and one green.

Transparents – One of each of the following: orange, yellow, green, blue, violet, red and brown.

There is no reason, of course, why you should not begin with a smaller selection. You can then add to your range of colours as you become more experienced and, inevitably, enthusiastic. Some enamel manufacturers in Great Britain supply inexpensive sample sets of enamels with a dozen or so colours in them. These are

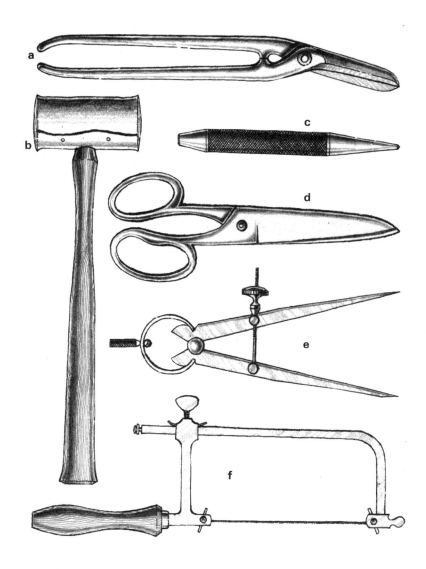

Fig. 9
a Duck-bill type metal shears
b Rawhide mallet
c Centre punch
d Scissors
e Spring dividers
f Jeweller's saw frame

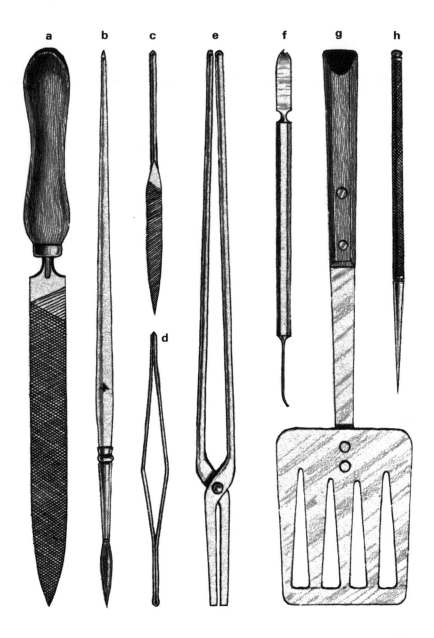

Fig. 10
a Handfile
b No. 4 sable paint brush
c Needle file
d Tweezers
e Metal tongs
f Stainless steel spatula
g Kitchen spatula
h Scriber

available in both transparent and opaque colours. Clearly the quantities supplied of each colour are small but these may well prove sufficient for your initial requirements.

Other materials and tools

The other items which must be obtained are as follows:

A bottle of enamel adhesive for securing the enamel during its application. This may be made up (page 28) or purchased from an enamel supplier.

A double-ended stainless steel spatula for the application of enamel, also obtainable from the enamel supplier.

Some sheets of medium coarse emery cloth.

A rawhide mallet and a medium coarse hand file with handle.

A pair of metal hand shears with cutting blades no smaller than 2 ins long.

A piece of soft asbestos approximately 20 ins square and $1/4$ in. thick.

Most of these items can be supplied by your local hardware store.

You will need a half-sheet of mild steel or, if your metal merchant can supply it, a half-sheet of enamelling steel would be better. 20 swg (gauge) is the ideal thickness for either of these although 18 or 22 swg (gauge) will also do. Your same local merchant may also be able to supply a piece of steel wire matting of $1/4$ in. weave for supporting your work in the kiln. If this is not available, a piece of 'expanded' sheet steel is a good substitute. Finally, from the local art supply store, a no. 8 sable paint brush and a sgraffito tool, although a steel scriber will be perfectly adequate.

None of these items is expensive and with all of them you will be well established. I have assumed that you already have such things as domestic storage jars, a pancake spatula, and scissors. Gradually as you become more experienced and ambitious you may well think you require more sophisticated equipment. These further items are not essential at the beginning and when they do arise they are not expensive. The most expensive item is the kiln, and since this is the key to the entire process it is very good value. The cost of the other things you will require will depend on the amount and range of enamelling you intend to do. It is worthwhile remembering at this stage that enamelling is an art craft which anyone can practise, regardless of the amount of experience he may or may not have had in other craft areas. Nonetheless, the purchase of good tools is always a sound investment. They will serve you better than cheaper ones and, provided they are not neglected, will last for many years. Ideally tools are kept at room temperature in a clean dry place. No matter how proficient you may become as an enamellist, remember your work can only be as good as you and your tools will allow. A complete list of all the items and their suppliers mentioned in this chapter appears at the end of this book (page 101).

3 Making the first piece of enamel

For this first exercise it is as well if the beginner starts with a simple piece of flat work. If a small kiln is available, a piece of ordinary mild steel or enamelling steel between 18 and 22 swg (gauge) and approximately 3 ins × 3 ins is a good starting size. These gauges of metal are not too thick for cutting and will heat rapidly without causing any cooling problems. The dimensions of the metal are based upon the average size of the firing chamber of the many small kilns available on the market. It is essential that your plate of metal can be put into the kiln without touching the sides, so the size of the plate to be used is governed by the size of the firing chamber. There must always be clearance between the edges of the plate of metal and the sides of the firing chamber. The plate must be put well to the rear of the kiln, because the opening and shutting of the kiln door cools the immediate front of the kiln and an uneven firing can occur if the work is too near the front.

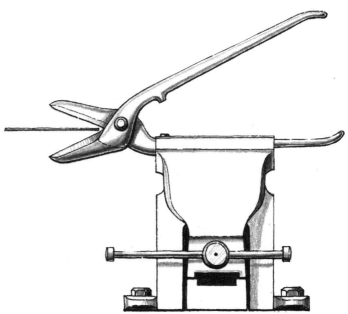

Fig. 11   Mount one handle of the shears in the vice. Open the blades fully and insert the metal as far into the throat as possible

Cutting the metal

Measure and mark out the metal with a ruler, set-square (triangle) and metal scriber. This may then be cut out with a pair of hand shears or metal snips (scissors). If you are working in a heavier gauge metal or do not have a strong grip it is very helpful if you mount one handle of the shears in a vice as illustrated (fig. 11); then, by a squeezing action of the hand, cut out the square of metal. This method gives a considerable mechanical advantage and allows quite thick gauges of metal to be cut quite easily in straight or curved lines. When cutting the metal, open the jaws of the shears fully and insert the metal right into the throat. Never cut right to the ends of the shears blades, as this causes a slight bend in the metal which will have to be removed by filing and hammering afterwards. The illustration shows clearly the correct and incorrect methods (fig. 12).

After you have cut out the square of metal, lightly file the edges with a smooth hand file. This is to remove any burrs or sharp points that have occurred while shearing. Check that your square of metal is flat. If not, tap it flat with a rawhide or wooden mallet. Do not use a hammer for this, as it will leave unwanted marks and creates tensions in the metal.

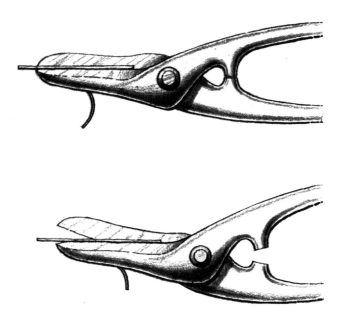

Fig. 12   Never cut to the end of the blades as in the top picture but to just before the end of the blades as in the second picture. This will give a clean and more continuous cut

Cleaning the metal

Hold the metal plate firmly on a flat surface and, with a medium coarse piece of emery cloth held in the fingers or wrapped around a block of wood (fig. 13a), rub one half of the surface of the metal clean and bright. It is essential now that you do not touch this clean part with your fingers; this would leave grease on the metal, which would form a barrier between the enamel and the metal and prevent any fusion between the two when the work is fired. Cover the half you have cleaned with either a thick piece of lintless cloth or clean paper and use this as your pressure point for holding the plate down while you clean the other half (fig. 13b). The success of your first piece of enamel depends largely upon cleanliness and this process must be carried out rigorously if you are to produce a satisfactory piece of work. Wash the plate under a running tap, again without touching it with your fingers. The metal can be held in any metal tongs or tweezers, although stainless steel ones are ideal. Dry the plate thoroughly and lay it upon a clean sheet of paper, and it is ready for the enamel to be applied. If for any reason you are unable to apply the enamel fairly soon after preparing the plate, submerge it in a vessel of clean water. This prevents any contamination or oxidation, and the plate will be ready for use again after it has been thoroughly dried.

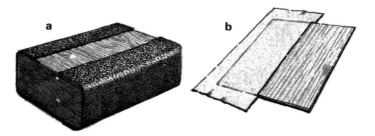

Fig. 13　**a** Wrap a piece of medium coarse emery around a piece of wood or a carpenter's sanding block
b Cover the half of the plate you have just cleaned with paper before cleaning the second half. This will prevent your fingers touching the clean area

Preparing the enamel

For your first piece of flat enamel I recommend the use of an opaque enamel. Translucent or transparent enamels are not generally suitable for use as background coats on mild steel. Normally enamellists use a specially prepared 'grip coat' on steel if they are enamelling any large amount; but for the beginner it is not necessary.

Already ground (that is, prepared) enamel straight from the maker often includes a small percentage of impurities. Usually this is enamel which has been ground too fine to stabilize in firing. To remove this, the enamel is washed. Place the enamel in a clean jar and cover with water. Stir this vigorously and the impurities will rise to the surface. Pour these impurities carefully away and repeat the operation until the water is clear (fig. 14).

Now place the damp enamel on a flat dish and dry it in a warm atmosphere. This may be done near the kiln but not in it. Over-rapid heating can cause the enamel to scatter and prolonged excessive heating will fuse the enamel into caked lumps. When dry, store the enamel in a clean screw-top jar. This is to prevent any impurities contaminating the enamel and affecting your finished result. Cleanliness is a basic requirement of the enamelling

Opposite
Fig. 15   The top of an ordinary screw-top jar may be converted into a sifting cap for the dusting technique, as shown here

Fig. 14   Cover the enamel with water, in a jar, and stir it. The fine dust will float to the top of the water while the good enamel will settle on the bottom of the jar

process and the skimping of this process or the cleaning of your metal will only mar your finished work. All enamels require washing to obtain their true brilliance. Opaque enamels usually require only one wash, but the transparent colours which rely upon complete clarity require at least five.

A container for the enamel

Undoubtedly the simplest way of applying enamel is by the dusting technique. This enables a large area to be covered quickly, smoothly and evenly. The best utensil for this is a small screw-topped jar. Remove the lid of the jar and drill or cut out with your snips (scissors) a hole in the top approximately one inch in diameter. Then cut a disc of fine wire mesh to fit just inside the lid. Now cut a cardboard or fibre washer with a hole slightly larger than the hole in the top of the lid. Assemble these with the mesh disc sandwiched between the inside of the lid and the washer (fig. 15). This is called a sifting top and there are ready-made ones available. Often, however, you will require a wider range of sizes than is available and it is simple to make your own. Into your sifting jar put a little more enamel than you think you will need to cover the metal to a depth the same thickness as the gauge of the metal.

Applying the enamel

On your first piece of enamel it is better if an adhesive is used to secure the enamel to the metal surface while it is being applied and fixed. When you are more familiar with the kiln and have a more confident and steady hand, you may consider the adhesive unnecessary on flat work. This adhesive is available as a ready-made fluid, or you can make it yourself. Its base is gum arabic or tragacanth, which may be bought from any drugstore or chemist. Tragacanth is best prepared overnight. First place the powdered tragacanth in an oven proof jar and cover with water, and leave it for about eight hours. Then simmer the mixture gently for about five minutes, slowly adding sufficient water to prevent the mixture burning. Allow this to cool and add more water, stirring until the mixture is of a more water-like consistency. This should be sieved through cheesecloth, or a material of a similar weave, into a storage jar. Mixed tragacanth is best kept refrigerated. It is inclined to lose its efficiency when stored at room temperature, and more so if near the kiln.

Fig. 16 Gently dust the enamel on to the metal plate

24267

Using a camel hair or sable brush, apply the adhesive to the surface of the metal. This should be done evenly and as thinly as possible for the best results. You are now ready to apply the first coat of enamel.

Lay a piece of clean paper, which should be bigger than your metal plate, on a flat surface near the kiln. Then, with a suitable large spatula tool, lift the plate of metal and place it on the paper, remembering not to touch the metal with your fingers. Any clean surplus enamel that falls on the paper can be returned to your storage jar after you have fixed the enamel, thus avoiding wastage.

Holding the dusting jar in your fingers at a slight angle, tap its sides lightly and dispense the enamel gently over the metal surface (fig. 16). Excessive shaking and tapping may deliver too much on one spot. Try to dispense the enamel as evenly as possible, without making ridges and valleys. As previously stated, the ideal layer is approximately the same depth as the gauge of the metal. However, particular attention should be paid to the edges and corners, where this thickness may be slightly exceeded. The enamel should now cover the surface of the metal completely, with only the thickness edge of the metal showing (fig. 17).

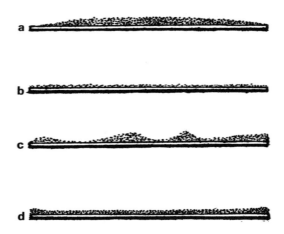

Fig. 17   Dusted enamel
a Wrong. The enamel is too thin at the sides and is heaped in the middle
b Wrong. The enamel is too thin all over
c Wrong. The enamel is applied too unevenly
d Correct. The enamel is applied evenly with slightly more at the edges

PRESCOTT HIGH SCHOOL LIBRARY
PRESCOTT, ARIZONA

Enamel applied too thickly may crack when cooling, and enamel laid too thinly may tend to burn away or leave a porous surface when fired. The application of enamel is quickly learned and in a short space of time you will achieve perfect results with this dusting technique.

Firing the enamel

Most enamels fire at around 750° C, 1382° F. When the metal and the enamel achieve this temperature the enamel ceases to be granular and melts into a semi-fluid layer over the surface of the metal, and a fusion takes place between the two. Most small kilns achieve this firing temperature in about twenty minutes, so you must ensure that your kiln has been switched on beforehand and is at the correct temperature when you are ready to use it. Small kilns do not have any of the expensive temperature controls. These are not strictly necessary, as you will soon learn the temperature of your kiln by looking inside it. As a very general guide 750° C or 1382° F appears as a bright red glow in the kiln.

To put the work in the kiln you will need a thick glove to protect your hand from the heat of the kiln, although this is not, thanks to the insulation of the kiln, as intense as you may imagine. You will also need something like a kitchen spatula with which to lift your work into the kiln. The work is placed in the kiln on a support, which may be made from any strong steel mesh or expanded steel sheet. Its purpose is to support the enamel plate as much as possible and at the same time allow the heat to reach the underside of the plate. To make a support, cut a piece of mesh with your shears, very slightly larger than your enamel plate. With a hammer or a mallet turn the edges over so that you have a flat surface to rest your plate upon and legs for the tray to stand on in the kiln (see fig. 30, page 44). There are ready-made stainless steel stilts available which support the work in the kiln, but I prefer with smaller work to use a steel mesh which supports it over a larger area, rather than balance it on the points of stilts.

With your spatula, lift the enamel plate carefully from the paper and place it on the support tray near the kiln. Then open the kiln door and lift the tray and plate and place them in the rear part of the kiln. Care should be taken not to jar the work against the sides of the kiln, as this will cause enamel to fall away. This will mark your work, and the fallen enamel will fire to the kiln floor and is very difficult to remove later. Carefully withdraw your spatula, leaving the tray and plate in the kiln, and gently close the kiln door. The enamel will take about three minutes only to fire. On some of the smaller kilns the firing may be seen through a peep-hole or window in the kiln door. If your kiln does not have one, you may open the door just sufficiently to see the stages of firing.

The enamel will gradually lose its granular appearance and assume a glossy surface. At this point you should remove the work from the kiln with your spatula, again taking care not to touch the sides of the kiln as you do so. The colour at this stage will bear no resemblance to the colour you applied originally, nor indeed to the true finished colour of the enamel when it has cooled. Care-fully place the tray and work upon an asbestos pad and close the kiln door. This asbestos pad should be in a warm draught-free and dry place, preferably near the kiln. Over-rapid cooling of the enamel will cause it to crack or shatter.

After ten minutes or so the fired work should be cool enough to handle, although it is still advisable to keep your fingers off the surface. If you wish to apply further enamel, the edges of the metal

Fig. 18   Lift the enamel plate with a kitchen spatula

plate should be cleaned, with a piece of emery cloth wrapped around a piece of wood, until they have a clean, bright finish. This removes the oxide (fire scale) which has developed during the firing. This scale can leave unnecessary black flecks in your enamel as it tends to fly during the cooling period and is difficult to remove at a later stage.

Depending upon how well you applied the enamel, you may notice areas where the enamel appears a little thin, possibly with parts of metal showing. You may apply further enamel to these areas, dusting it on as before, and fire the work again as just described. You may also notice that the enamel is not as thick as when you applied it. This is because the grains have fused with each other and there are no longer the many tiny spaces between them. If you are going to apply further enamel, it is important that the fired thickness of enamel should not exceed the gauge of the metal it is fused to, as this may cause cooling and contraction problems.

Fig. 19 One coat only of enamel, applied very thickly in places to obtain a strong texture. An early exercise by a mature student

Applying a design

With this first piece of work it is best to keep to simple or geometric forms. A successful and easy way of doing this is by the stencil method. Choose a piece of thin, strong cardboard (cereal packets are ideal for this) and with a pair of scissors cut out a square the same size as your enamel plate. On to this draw with a pencil a smaller square and cut out this square (fig. 20a). Lay a piece of clean paper on your flat surface near the kiln, with your enamelled plate upon it. Over your plate lay your stencil, and inside the cut out part, where the newly fired enamel is showing, paint adhesive thinly and evenly. Select any colour you think will look attractive with the base coat and put sufficient of this into your dusting bottle to cover the exposed area. Dust the enamel onto this area, taking care to disperse it evenly without making the hills and valleys previously mentioned. Do not worry at this stage about the

Fig. 20
a Cut a square out of a piece of cardboard and lay it over the centre of the plate and dust enamel into the space
b Carefully remove the cardboard and brush away the surplus enamel with a paint brush. Fire this
c Lay the cardboard at an angle over the fired square
d Dust a layer of transparent enamel into the space, remove cardboard and fire as before

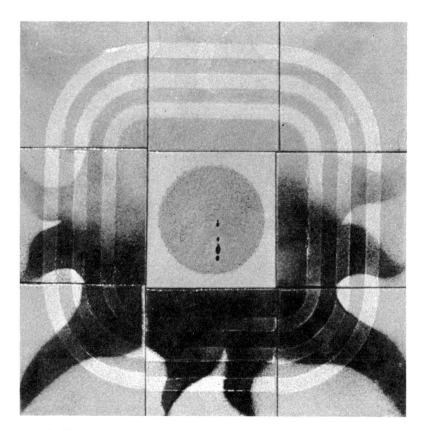

Fig. 21

Figs. 21 and 22   Enamelled steel panels by 1st-year students at Somerset College of Art, England, showing a variety of coarse, smooth and glazed firings. The enamels were applied by the dusting and stencil methods with the occasional use of a spatula (see also figs 3 and 4) surplus dispensed upon the edges of the cardboard stencil. When you have finished applying the enamel, carefully support the edges of the card and lift it away from the enamelled plate, taking care not to drop any surplus upon the plate underneath. If you do so, this may be removed by using a clean, soft, dry sable brush, sweeping away from the edges of the enamel pattern you have applied (fig. 20b).

With your spatula lift the plate on to the support tray and fire it exactly as before, and allow it to cool slowly. You should now have a perfect square reproduction of your stencil. You may now put the surplus enamel, and also any on the clean paper which was underneath the plate, back into your storage jar for use again at a later time. If you wish, you can use the same stencil again but at a different angle (fig. 20c and d). A transparent enamel may be used provided the previous colours are light enough to allow the transparent enamel to be seen. Whichever colour you choose, apply it as described above, then fire and cool it as before. Further enamel may yet be added but I would recommend that for this first piece on steel you do not fire it more than five times.

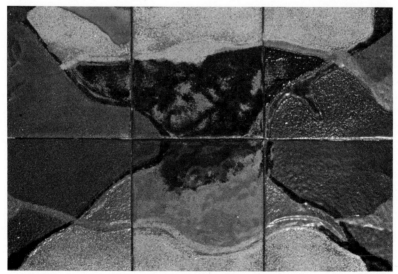

Fig. 22

Fig. 23   Enamelled steel panel by Karl Drerup, USA

After you have practised this dusting technique and are able to fire a reasonably smooth ground coat of enamel, you may wish to try another method of applying a design, known as sgraffito. This is simple and effective and is described on page 68. The design on steel tiles shown in fig. 1 (frontispiece) combines the dusting and stencil technique with spatula work. The antlers were applied in thick transparent enamel which, when fired, became semi-opaque and opaque in the thicker areas. At all times it is important not to touch the surface of the metal or enamel with your fingers and to remove the fire scale from the edges of the metal plate after each firing. Keep your work place clean and tidy and approach each stage methodically, and your enamelling will be successful. All the processes so far described are simple and straightforward and with practice are accomplished in very little time. Perhaps one of the many attractions of enamel is that a pleasing and permanent result can be achieved so quickly.

Fig. 24   Dish with enamel freely applied. Mature student's early exercise, Blue School, Wells, Somerset

4 Making a piece of low relief enamel

For the process described in this chapter it is best to consider the use of gilding metal or copper as an alternative to steel for enamelling upon. Having successfully enamelled a flat plate, the next step is to undertake a piece of low relief work. A small tray is an ideal subject for this.

Preparing the metal

Take a piece of either copper or gilding metal, and mark out and cut a rectangle 3 ins × 2 ins. Use the metal cutting method described in chapter 3 if you wish, although you will find both of these metals much softer and easier to cut than the steel. With a ruler and a set-square (triangle), lightly scribe a rectangle $^3/_4$ in. × $^1/_2$ in. inside the edges of your piece of metal. From the corners of your inside line scribe a firm triangle to the outside corners of the metal plate as illustrated (fig. 25a). With the shears, cut down from the outer edges to the inside corners of the scribed lines, thus cutting out the small triangles at the corners (fig. 25b). Take care not to cut beyond this point. Open the shears blades fully and do not cut to the end of the shears. If the metal has distorted slightly during the cutting, tap it flat with a rawhide mallet before going any further. You will now need to prepare a piece of wood 1 $^1/_2$ ins × 1 in. and approximately 3 ins deep. Mount this in a vice with the 1 $^1/_2$ ins × 1 in. surface protruding about 1 inch from the jaws of the vice. Place your plate of metal, with the scribed marks facing downwards, on top of the piece of wood. If you have marked out and cut your metal and wood accurately, the scribed lines will correspond with the edges of the wood. Now, with the rawhide mallet, gently tap the sides of your tray over to an angle not exceeding 45° as illustrated (fig. 25c). Take care not to bend any one part too quickly. Try to bend the metal along its edges evenly and gently. Do not attempt to get a sharp corner at the base of the tray as this is almost impossible by this method. In any case, enamel is difficult to fire in sharp corners. A small radius at corners is always preferable. Ensure that your tray base is flat before going any further, by tapping it with the mallet on the wooden former. All this should be done quite quickly and with the minimum amount of malleting. Your tray is now formed and is ready for the next stage.

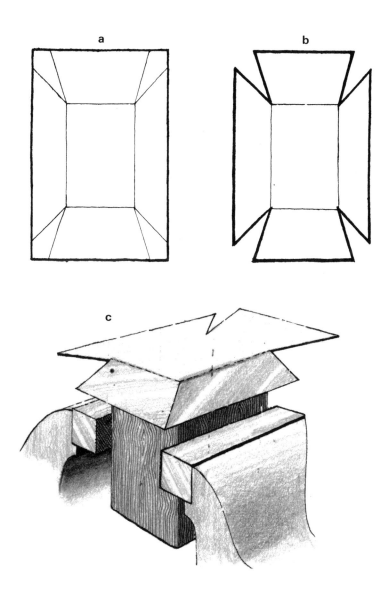

Fig. 25   Cutting the metal
a Mark the inner rectangle and corners with a scriber
b Cut the corner pieces away with shears or a jeweller's saw
c With a rawhide mallet, tap the sides of the tray over on a piece of wood mounted in a vice

Cleaning the metal

These metals may be cleaned by the same methods as described in the previous chapter for steel. However, a better method is suitable for non-ferrous metals such as copper and gilding metal and is known as pickling. Before pickling, the work should be warmed in the kiln for not more than one minute to remove any grease that may be on the metal. It should be removed from the kiln and plunged instantly into cold water.

When cold the metal can be put into the pickle, which should be contained in an acid-resistant plastic or ovenware glass bowl. The pickle can be of one of two types, and you may choose which of the two you find the easiest to obtain or store. I prefer to use the first type, which is made by pouring three parts water to one part nitric acid into the bowl. Always remember when making this solution to add acid to water (fig. 26). Never add water to acid: this causes a violent chemical reaction and a severe personal injury may result. If you can only obtain commercial nitric acid then the mixture should be two parts water to one part commercial nitric acid.

Fig. 26   Always add acid to water, NEVER add water to acid

The second type of pickle is made in the same sort of bowl. Put one pint of kitchen vinegar into the bowl and add to it one egg cup of cooking salt. Stir in the salt until it has dissolved. This vinegar pickle has a shorter life than the acid, remaining effective for only two or three days.

When preparing either of these solutions, the bowl should not be filled. Sufficient room must be left for the work to be immersed in the pickle without displacing any of the solution over the sides of the bowl.

The work should be immersed in either of these solutions for about ten minutes, after which time it should appear pinkish in colour. Do not leave the work in the pickle any longer than is necessary, or the pickle solution will begin to erode the surface of the metal and will leave it too thin for practical purposes. The work may then be removed with brass tongs or stainless steel tweezers and rinsed carefully under the tap (fig. 27). Care should be taken doing this. Run the tap slowly, as any solution splashed by the force of the water on to your clothes will slowly burn holes

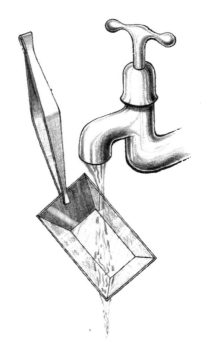

Fig. 27 Holding the tray in tweezers, rinse the surplus acid away under a slow-running tap.

in them. If you have a plastic or rubber apron it is as well to wear it when dealing with the pickle at any stage. After rinsing, the work should be scoured with steel wool or scrubbing powder and rinsed clean again. It may then be dried with lint-free cloth. The tray may now be lifted with the tongs or tweezers and placed upside down on a clean piece of paper on a flat surface near the kiln. Once again, take care not to touch the metal with your fingers as this will entail repeating the cleaning process again. The metal tray is now ready for applying the enamel.

Counter-enamelling

Because this piece of work is in relief it is necessary to counter-enamel it, that is, to enamel the back as well as the front. When you first fired the steel plate you may have noticed a slight curve appear in the metal after it had cooled. This is because enamel cools and contracts at a faster rate than metal and, as it has now fused to the metal, the enamel bends the metal as it cools (fig. 28). Further applications of enamel would increase this problem until the metal could bend no further and at this point the enamel would begin to crack and shatter away from the metal. By counter-enamelling you are counteracting this effect by placing a pulling force on the other side of the metal, thus keeping it flat.

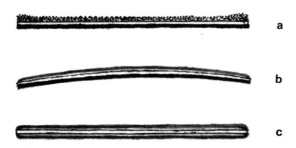

a

b

c

Fig. 28
a A flat metal plate enamelled on one side only, before firing
b After firing, the plate has curved
c When counter-enamelled and fired, the plate is level

Counter-enamel is usually made from the surplus enamel from previous applications. Despite this, it has a very attractive speckled finish. Often the counter-enamel is more attractive than the so-called finished side. You may not yet have sufficient surplus enamel to make a counter-enamel coat, in which case you will choose a colour which you think will complement the other side of your tray. Having prepared the enamel, place the tray upside down on a piece of cardboard and paint adhesive over the back of the sides of the tray, leaving the base clear. Then dust enamel over the adhesive, endeavouring to get as even a deposit as you can. If you do sprinkle enamel on the clear base this may be brushed away with the sable paint brush (fig. 29). Carefully, with tweezers at opposite corners of the unenamelled base, lift your tray, turn it over and stand it on its base on a clean piece of cardboard. The interior may now be painted with adhesive and your second choice of colour dusted on to it. Now carefully lift the card under

PRESCOTT HIGH SCHOOL LIBRARY
PRESCOTT, ARIZONA

Fig. 29   Brush away the surplus enamel from the underside of the base of the tray with a paint brush

the tray and place it on the wire mesh support. Slowly slide the cardboard from under the enamelled tray until it is standing on the wire mesh. It may help to push the point of a scriber against the edge of the metal tray as you withdraw the card, in order to ease it off (fig. 30). Lift the support and tray with your kitchen spatula and place them in the rear part of the kiln. Fire and cool them in the usual way. When sufficiently cool to handle you may find certain areas which did not take the enamel as well as others, in which case you apply further adhesive and enamel, and refire the tray. You may well consider using another colour at this stage. By carefully dusting the upper edges with this other colour you will be able gradually to 'mist' this almost imperceptibly into your original colour. This is a most attractive method and is well worth trying at this stage. There is no reason why you should not try another stencil pattern on some part of your tray, but do not over-decorate this small piece of work. Between each firing, as with the steel plate, you must clean the edges of your tray to prevent the fire scale marking your enamel.

Fig. 30   Slowly withdraw the card from under the tray. The scriber may be used to prevent the tray moving with the card

Finishing the enamelled tray

The edges may now be finally cut back to a sharp and crisp edge with a piece of emery cloth wrapped around a piece of wood. This will enhance the appearance of your work considerably. If there is any enamel on these edges it should be cut back with a fine carborundum stone; enamel is extremely hard and will merely wear away your emery cloth. After using the stone, finish the edge with emery cloth. The underside of the base of the tray may be rubbed with emery cloth and finished with a fine grade of steel wool. This area of metal will in time oxidize, particularly if kept in a warm place. This may be prevented by applying one of the many transparent lacquers or celluloses that are available. It should be applied as quickly and as thinly as possible, avoiding tears or runs, when the base is absolutely clean and dry. It will have no ill effect whatsoever upon the enamels.

You have now completed one of the more demanding aspects of enamelling and may well consider producing another similar piece in order to practice what you have learned so far. If you wish to change the shape of the piece, do not design too deep a form. This will create problems you are not yet able to manage. Keep the form in low relief and try further dusting variations. Alternatively, try different stencil patterns. You will find it better not to use too many colours on a small piece of work as they will tend to conflict with each other. Generally speaking, one bright colour is enough on a small piece when used in conjunction with other quiet ones. Red is an expensive colour in enamel and is particularly volatile in its visual effect upon other colours, particularly on small pieces of work. A few fortunate people are gifted with what is known as a colour sense. Most of us have to develop this elusive quality and learn from the experience gained from our mistakes. A colour sense can be developed, and it is possible that the next chapter will be of help towards this. Certainly enamelling will develop your colour sense, but it is a very special type of colour. The quality of enamel colour is quite exclusive, but is nonetheless something which can be appreciated by everyone.

Fig. 31 *Mayflower*. Enamelled gold and copper by Geoffrey Franklin. The vessel is in gold with small spots of enamel added by the spatula and cloisonné methods. The copper background is in low relief with the enamel applied by dusting and 'brushed' while dry to obtain the effect of sea. Approx. 7 1/2 ins. square

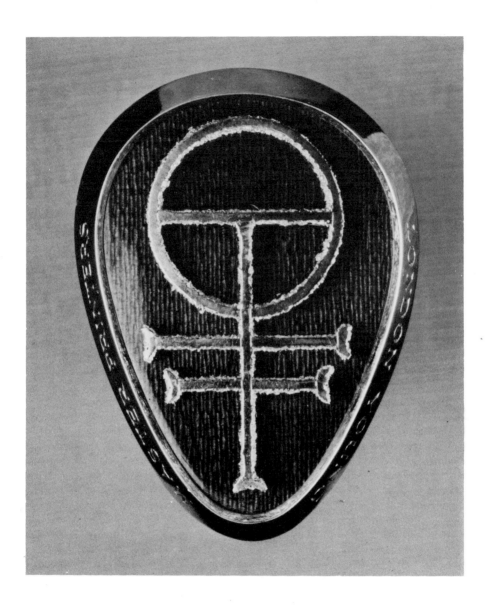

Fig. 32   Enamelled badge of office by John Donald, England, for the London Young Master Printers, in yellow and white gold with red enamel

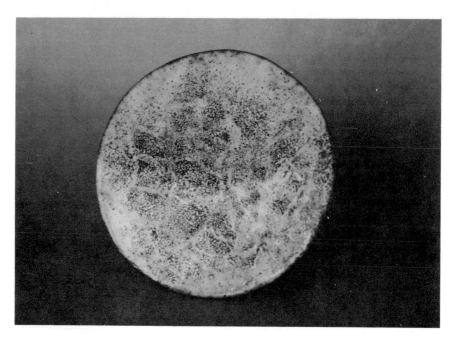

Figs 33 and 34   Dishes, base coated, with subsequent enamel applied through a hair net. Schoolboy aged 12, Wells, Somerset

Fig. 35 Dish with edges 'repaired' after enamel had shattered away. This does not detract from the design. Schoolboy aged 12, Wells, Somerset

5 Test firings: an essential practice

If a reasonable degree of control of the finished piece of enamel is to be achieved, it is essential that test firings be made of every individual colour used.

Enamel manufacturers each have different names for their individual colours, rather as paint manufacturers have. You will find it much easier to ignore these names and use only the manufacturer's reference number. More often than not the names will not agree with your interpretation of the actual colour, and when a number of enamels are being used these names can be confusing. Usually the unfired enamel bears little resemblance to the fired enamel colour or its manufacturer's name.

I have found the simplest way of referring to opaque enamels is by making a set of 22 swg (gauge) mild steel plates, $1\,^1/_2$ ins square, drilling a hole in the top centre of each of these, and punching the number of an enamel onto the plate immediately beneath the hole. I then apply the enamel by the dusting technique on the remaining surface of the metal, and fire it. Having made a number of these, it is advisable to prepare a wooden board and screw cup hooks into it at roughly 2-inch intervals on which to hang them (fig. 36). You will then have a permanent and variable colour chart which will allow you to detach colours and lay them beside each other when making up your colour schemes.

Opaque firing variations

Having established that the colour of unfired enamel is quite different from that of fired enamel, the next step is to appreciate the variation of depth, tone and texture that is possible within each colour. This is achieved by the degree of firing. Generally speaking, most people tend to overfire their enamels, believing that when the high gloss appears the enamel is truly fired and fused to the metal. This is basically correct, but enamel may also be fired and fused to the metal well before this gloss appears.

Opaque enamel for general purposes may be fired to three different degrees of finish: first, at a low temperature when the enamel is only just fused to the surface of the metal and the enamel retains a coarse, gritty or sand-like appearance; secondly, a medium temperature firing, whereby the enamel loses its gritty appearance but does not achieve a high gloss (this is sometimes called an eggshell finish); thirdly, a high temperature firing, when

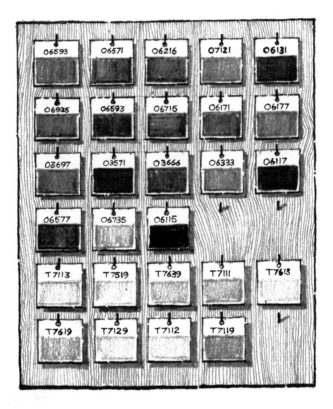

Fig. 36 Test plates for opaque and transparent enamels. Each has the manufacturer's number punched on the metal before applying the enamel. A wooden board is used to hang the test plates on. This is also very useful as a reference chart

the maximum glaze is obtained. There are clearly variations to be obtained within these three basic firings, all of which are achieved by the length of time the enamel is left in the kiln. Depending upon the size of the piece of enamel and temperature of the kiln, a low temperature firing for the coarse finish may only need to be in the kiln for thirty seconds or so. A full glaze would be over a minute.

For my own convenience I prepare three test plates, one for each of the finishes mentioned for each colour of enamel; and I would recommend you to follow this or a similar practice. The basic colour does not vary very much but the variation in tone and texture is quite dramatic.

A point to remember, when planning schemes for a piece of work, is that the high firings must be made first and the low temperature ones last. Obviously if this procedure is reversed a beautiful sand-like texture will glaze over with subsequent high temperature firings. If you do fall into this trap, and nearly everyone does at some time, this can be remedied by applying further enamel to the area originally intended for a coarse finish, and refiring it at a low temperature. You may find this coarse texture a little difficult to obtain at first, since the length of time that the piece of work is in the kiln is critical.

As a general practice it is better to err towards the underfired rather then the overfired. Underfired enamel should have melted just sufficiently to have adhered to the surface of the metal. This may be tested by scraping gently the surface of the enamel with a scriber or the point of a sgraffito tool. If the enamel has not fused sufficiently it will scrape away. Properly fired enamel will not scrape away. If it has not fused sufficiently, replace the work in the kiln for a few more seconds and test it again after removing it from the kiln. This can be repeated until the enamel is fused to the metal but still has a coarse appearance. It is quite impossible to underfire enamel once it has been glazed. This can only be achieved by applying new enamel.

Transparent enamels

The tonal and textural firing variations described above are purely for use with the range of opaque enamels. Transparent enamels must without exception be glazed to achieve their full transparency. The underfiring of these results in a dull, cloudy appear-

Fig. 37   Cloisonné relief panel by Karl Drerup, USA

ance which, except for an unusual effect, would be unacceptable. Also, as mentioned in chapter 3, transparents lose their effect when applied directly onto mild steel. Test firings for these should be made on a non-ferrous metal such as gilding metal or copper or, if applied on mild steel, over the top of a white opaque ground-coat.

Further tests

An essential extension of this reference chart is test firing for any of the variations of technique and colour that you as an enamellist may wish to try. The effects of overlaying one transparent colour over another, the effect that gold and silver have upon the colour of enamel, particularly the transparents, should all be tested and experienced before being used on a finished piece of work if you are to predict and control the finished appearance. Furthermore, a notebook will prove an essential record of methods and colours used. If you are using a three-colour sequence, write the enamel reference numbers and the sequence down as you test fire them. This applies equally to colour mixes. If, for example, you are using green and yellow, write down the proportions of each and their number in a record book. All too often an enamellist has to leave his work, and finds on returning that he is unable to remember how or with what he achieved that particular result.

Enamels also vary in hardness, that is, the temperature at which they fire. This may be established by placing a group of enamels in small quantities simultaneously in the kiln and watch-ing them fire. Note the order in which they melt, in your notebook.

Test firing and method records are an essential part of the process of enamelling. In no way does this destroy the mystery or magic of the craft. They should be regarded purely as a means to an end, a necessary experiment to achieve an exciting, individ-ual result. On page 98 you will find some notes on the behaviour of different enamels colours.

Fig. 38   Dish with a very fine second coat of enamel. Schoolboy aged 11, Wells, Somerset

6 Enamelling in the third dimension

With this third exercise I will describe the enamelling of a silver table-napkin ring. This ring will have no seam in it, having been cut from a piece of seamless silver tube 1 $^1/_2$ ins in diameter, and is 1 in. wide. Metal of this sort is available from any precious metal dealer; perhaps a little expensive, but probably not as much as you would imagine. A dealer will give you an indication of the cost.

You will find silver a beautiful metal to work with. This is one of the many reasons it was popular in earlier times. It will respond quickly to working and polishes readily. Its disadvantage is that it is soft and will not withstand harsh treatment. Every enamellist should work with silver at least once, and for a small piece of metal purchased directly from the bullion dealer the cost is not high. Enamelling (that is, fine) silver should be asked for, as this has a slightly higher melting point than other types of silver and also allows more brilliance.

Enamelling a ring of this kind involves many of the techniques so far described, plus one or two others. You need not obtain silver to follow this exercise. It may be carried out with either copper or gilding metal, with the exception of the paragraph specifying a special pickle for silver. A serviette ring is chosen for this exercise because the ring involves processes which are typical to the general three-dimensional field of enamelling. Many people will develop their own methods of application and each different project will require a variation of approach.

Cleaning the table-napkin ring

First polish the outside and inside surfaces with a fine-grade steel wool to as high a finish as you can obtain. Then gently warm the ring in the kiln to remove any grease on it, as described on page 40. It should then be pickled.

The pickle for silver is made up of four parts water to one part sulphuric acid. It should be heated, not boiled, in a glass oven-ware container. The silver will soon assume a chalky white appearance. It is then removed from the pickle, rinsed carefully under the tap avoiding splashing, and dipped in a finishing solution of one part nitric acid to one part water. It is then rinsed again under the tap and dried with a lintless cloth. At no time should the work be touched with iron or steel tongs or tweezers when it has acid on it. This will cause a copper plate to develop on the surface of the silver which will have to be removed. Use only brass or copper tongs. When pickled, carefully lay your ring on a piece of clean paper, edge downwards.

Applying enamel to the outside of the ring

Now select two transparent enamels which you think are compatible. Transparent blues and greens look particularly attractive on silver, as do the reds on copper and gilding metal. Paint the outside surface of the ring with adhesive. Place a quantity of one of the enamels in a saucer and thoroughly impregnate it with wet adhesive. This enamel should be mixed thoroughly with the adhesive to a cream-like consistency, but must not be too fluid. Take your stainless steel spatula and gather a small quantity of enamel on one end of it, and imbed this against the upper part of the outside surface of the ring. Continue this process around the ring until a band of enamel $1/4$ in. deep has been made (fig. 39a). The enamel will remain pliable for some time, during which you should attempt to flatten it against the ring in as thin, even and straight a layer as you can. If the enamel tends to dry out during this time it should be dampened with more adhesive, by dipping the end of your spatula in the adhesive bottle and dabbing it upon the applied enamel. Attention should be paid particularly to the

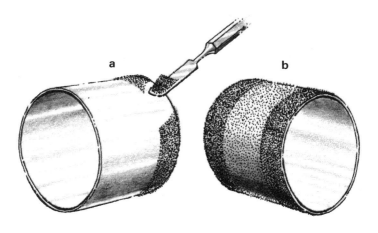

Fig. 39.
a Apply the enamel impregnated with adhesive to the outside edges of the ring first, with a spatula
b Fill the centre panel last

extreme edge of the metal. This is where the enamel is at its weakest and the thin even layer may be slightly exceeded here. Do not flood the enamel with adhesive, as this will tend to cause the enamel to run and will spoil the straight edge you have made. If you find difficulty in doing this with the ring lying down, you can pick it up with your fingers, touching the inside surface only. On no account touch the outside surface.

When you have completed the top band, repeat the process on the lower outside surface, so that you are left with a clear band of metal half an inch wide in the centre, running around the napkin ring. Take your second enamel and mix it with the adhesive as described. Apply this in the same way to the half-inch strip which has been left bare in the centre of the ring, trying to keep the line as straight as possible (fig. 39b). If the enamel is of the right consistency this will not be too difficult. You will, of course, improve with practice in this technique. Be sure that the enamel is as flat as you can get it and that it is not too thick.

Firing the enamel

Place the ring gently on the wire mesh support, edge downwards, and lift this with your kitchen spatula. Open the kiln door and warm the enamel near the entrance to the kiln, and then remove it, closing the kiln door immediately. It is essential that enamel be perfectly dry when it enters the kiln. Check that it is perfectly dry by looking to see if any steam is rising from the enamel. This may have to be done several times before you are satisfied that the enamel is quite dry. If the enamel were put in the kiln damp it would fall off the ring as the moisture in the enamel expanded into steam.

When you are quite sure there is no water left, place the wire tray and ring to the rear part of the kiln and close the door for a few seconds. When firing silver you must be very careful not to overfire the enamel. This can result in your melting the metal. Even enamelling silver has a very much lower melting point than copper, gilding metal and certainly steel. You cannot therefore neglect the timing of the firing as this is critical. Watch the enamel in the kiln carefully, and as soon as a glaze appears remove the work immediately.

If you have been a little too quick you may return the work to complete the firing. You should also cool silver a little more slowly than the other metals. You should now apply any further enamel to the outside that may be needed to refine the surface. After this dry, fire, cool slowly, pickle and rinse the ring again ready for the next process. Do not leave the ring in the pickle too long; it should be immersed just long enough to remove the black fire stain. If it is left in too long, the pickle will begin to attack the enamel. You should only handle the outside of the ring now as it is the inside upon which you intend to apply the enamel.

Applying enamel to the inside of the ring

For this I would recommend the use of one of the two colours used on the outside. Mix the enamel as described and apply it by the spatula method, attempting to obtain as thin and as even a layer as possible. Pay particular attention to the edges, keeping the enamel moist and pliable with adhesive as necessary. For the inside it would be better to put one colour right over. This should not conflict in any way with the outside finish. Dry this thoroughly, fire and cool it. Should the inside require any further enamel this should be applied and fired.

Finishing the enamel

The outside edges of the ring may now be rubbed flat on a piece of emery cloth on a flat surface. Should there be any enamel fired to the actual outer edge, this should be removed by rubbing with a damp carborundum stone first. Those who seek perfection will have noticed an undulation in the surface of the enamel. This surface may be perfected by rubbing it flat with the carborundum stone soaked in water. Care should be taken not to cut right through the thin enamel layer to the metal underneath. The ring should be rinsed thoroughly under the tap to remove all traces of carborundum and then dried. The enamel will now have a rough coarse appearance, and it should be finally fired to remove this texture and to give the enamel its fire glazed effect. This last process may well not be necessary or desirable for your purposes.

The use of transparents is one of the most exciting and rewarding aspects of the craft of enamelling, provided the metal underneath has been prepared adequately and the enamels have been thoroughly washed and applied thinly. These comments apply also, of course, to the opaques. The biggest single problem facing the enamellist working in the third dimension is making the enamel stay where you wish to put it. The secret lies in mixing the enamel and the adhesive to the correct consistency. Clearly you will improve with experience and practice but you should always approach each problem remembering that the method of application used for the previous piece of work may not be suitable for the next one.

7 Enamelled jewelry

Neither in ancient nor modern times has jewelry depended solely upon the use of exotic and precious stones for its beauty. Recent years have shown an even more diverse use of non-precious metals and of semi- and non-precious stones. Enamel has played an important role in this, since much of its colour imitates and enhances these precious materials. It is also a superb jewelry medium in its own right and when used appropriately can enrich the appearance of its wearer. The most-used enamelling method for jewelry is cloisonné. The technique consists of laying thin wires on their edges on a plate of metal, to act as separators between the various colours of enamel used. In so doing, the wires act as the lines of the pattern or as a coloured drawing and become an integral part of the coloured decoration. The word 'cloisonné' is derived from the French and means 'partitioned areas'.

To make a piece of cloisonné enamel

Cut a copper or gilding metal disc with a diameter of approximately 2 ins. Cut each side clean, bright and rough with emery cloth to ensure a good surface for the enamel to fuse to. Counter-enamel one side, and with a transparent flux or white, enamel the other. Transparent flux is applied and fired exactly as described for the coloured transparent enamels in the previous chapter.

Take a suitable length of very thin wire, preferably of the same metal as the disc. This wire should be as near as possible 30 gauge thick and 18 gauge high. Bend this with tweezers or small pliers into a pattern which will allow the wire to stand on its edge without falling over. It should also fit inside the diameter of your disc as in fig. 40b. Make three or four more such patterns, each time reducing the diameter so that the wires will fit inside each other and leave room for enamel in between. Do not bend the wires too sharply or you may experience difficulty packing the enamel into these small spaces later. Butt joint the wires as accurately as you can where the ends meet each other. Do not attempt to silver solder these unless you have previous experience of such work.

Paint the undersides of the wires with adhesive and place them on your enamelled disc, with the counter-enamel at the back. Make sure that you have joined the ends of the wires as best you can and then fire it in the kiln. This will fix the wires firmly to the enamel. If you find a wire which has risen slightly, it may be gently pressed down with your tweezers or spatula while the metal and enamel are still hot as you remove it from the kiln. Some enamellists claim that standard ground enamel as supplied by the manufacturer is too coarse for cloisonné purposes. If you have not made any of the bends too sharp in the wire, or put any of the wires too close together, you will not find this is so. If you wish to use a finer grade, it can be specially purchased from the manufacturer at quite a high relative cost, or you can grind it yourself to the grade of a fine table salt in a pestle and mortar.

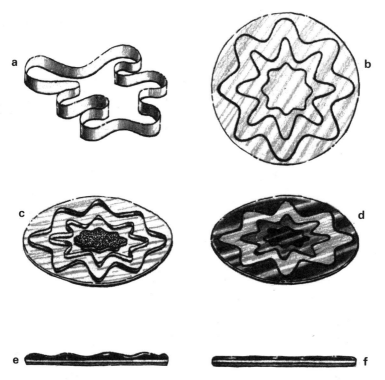

Fig. 40
a A typical free-standing cloisonné wire pattern
b The exercise pattern laid on the plate
c The centre cloisonné bed filled with enamel
d The entire pattern filled and fired
e Side view of the plate after firing but before stoning
f Side view of the plate after stoning flat with a carborundum stone

Dampen a small quantity of enamel with water and with your spatula apply this to the centre wire pattern on the disc (fig. 40c). You will find it helpful to use the point of a scriber as an aid while doing this. It is important to push the enamel firmly into the small corners of your design and to pack the enamel down as tightly as you can. When this is done apply the second colour to the adjoining bed, and so on until the disc is covered with the enamels of your choice. Try to avoid leaving grains of different colours on the various beds, and ensure that the enamel is built up to the tops of the cloisons (wires). Dry the enamels as described in chapter 6, and fire.

You will almost certainly find that the enamel has sunk below the level of the wires. In this case, pickle the disc to remove the fire scale, and apply more enamel, this time fractionally higher than the wires, and then fire again. Now with a wet, medium carborundum stone, rub back the enamel until all the wires are showing. Then use a fine carborundum stone to finish this process. The finer stone will remove any burrs on the wires caused by the coarser stone and will leave the enamel surface smoother. The wires should now appear clean, bright and crisp and your enamel should have no undulations (fig. 40f). The colours, although a little dull, should be clearly defined. If any of the beds are not completely full, further enamel should be added, fired, and the disc restored accordingly. The disc should now be fire glazed to achieve its full enamelled finish.

An extra feature is to polish the edges of the wires and disc with a jeweller's felt bob (buffing wheel) on a polishing motor, using jeweller's rouge as an abrasive. If you do not have these, however, your work will be satisfactory without this added professional touch.

The essence of designing for cloisonné is to use the wires not merely as separators but to achieve a flow and rhythmn with them. Your designs may be prepared on a piece of identical sized paper and the approximate lengths of your wires ascertained by tracing the pencil lines with a piece of string. Colours may be merged and intermixed within the beds, and areas can be left without enamel. Again avoid clashing colours in such close proximity to each other. It is worth while considering yourself as the wearer for these pieces and you will know best which colours do or do not suit you.

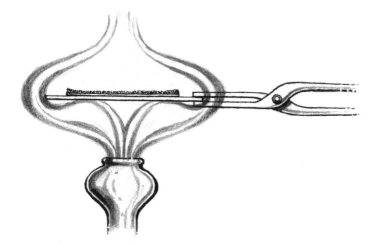

Fig. 41 The gas torch is held under the support plate, which is held with tongs. The flame then surrounds the enamel but does not touch it.

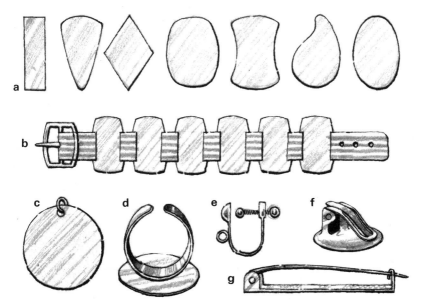

Fig. 42 Manufactured pieces for making jewelry:
a A selection of blank shapes suitable for enamelling upon
b A bracelet made from similar pieces; each piece has two thin slots pierced in the sides through which a wrist strap is threaded to make the bracelet
c A circular pendant blank with ring for cord or chain
d A ring
e An ear screw
f An ear clip
g A brooch pin

Firing with a gas torch

Small pieces of enamelled jewelry need not always be fired in a kiln. Enamel may be fired in the end of a gas torch, provided the enamel is not bathed in the flame, only in the heat of the flame. This is done by placing a flat disc of metal under the piece of work to be fired. The gas torch is held underneath this with the flame firing upwards so that it hits the underside of the protecting disc of metal. This distributes the flame around the enamelled work but not upon it. The work is thus encased in heat but not in the flame. By this method, you can observe easily how well your enamels are fired, and it is particularly suited to small test firings. The size of work is clearly limited to the size and power of the gas torch.

There are many jeweller's *findings,* as they are called, which will assist you in your enamelled jewelry work. Some are available from the enamel supplier and these include the prepared blanks for jewelry pieces. There are different shapes and sizes of rings, buckles and clasps, ring shanks, cuff-links, cuff-link blanks and springs, chains with clasps, buttons and brooch pins. You may cement these and other attachments with any of the modern resin adhesives to the backs of your own enamelled pieces. The cloisonné disc described in this chapter will make an excellent brooch by this method. Some enamel suppliers or craft supply houses (and bullion dealers in Great Britain) will also supply earring attachments, bracelet catches, a wider selection of chains and clasps, etc. Suppliers will send a catalogue of their products on request.

In addition to these you can also use wristwatch straps and leather thong in a variety of ways for your enamelled units. I have

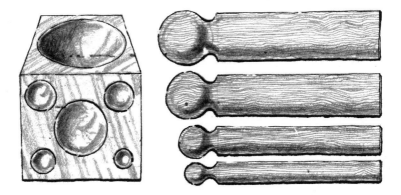

Fig. 43   A doming block and wooden punches

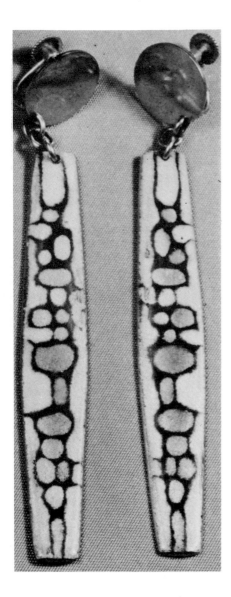

illustrated some simple jewelry forms you may adapt for use with them (fig. 42).

For those readers who wish to make completely original forms an essential piece of equipment is a doming block and a set of punches. The block is often made of brass and each punch is slightly smaller than its parent hollow in the block. A disc of metal is cut out slightly larger than the hollow selected and is annealed, that is, made soft by heating it to a dull red. The disc is placed over the hollow and the appropriate punch is put in the centre of the disc and hammered into the hollow, making a half-round cup shape. Small cups and hollows of all shapes can be made in this manner. This offers a wide range of forms for enamelled jewelry purposes.

Once again I would advise novices to avoid soldering and use the very strong modern adhesives wherever possible for cementing on the various attachments such as pins' and clips. Soldering on a jeweller's scale is a particularly specialized craft process. To complicate this with enamel is asking a lot of a beginner's ability.

Fig. 44 Earrings by Gerda Flockinger, England. Here the enamel is applied with the spatula and then glazed sufficiently to allow the edges to bleed very slightly

Fig. 45 Enamelled copper brooches and pendants by Madge Martin of Canter-
bury, England. These are simple but excellent examples of trail enamels (see page
75) with transparents and opaques used together

Sgraffito

Cloisonné is, of course, not the only method of enamelling jewelry.
You may also enamel a background and then apply a further coat
of a different colour. Before you fire this second coat, scratch a
pattern through the unfired enamel with a scriber so that the
glazed enamel background shows through (fig. 48). Then fire

Fig. 46   Brooch by Madge Martin (see also fig. 45)

PRESCOTT HIGH SCHOOL LIBRARY
PRESCOTT, ARIZONA

Fig. 47   A pin dish, using the sgraffito technique, made by a schoolboy aged 11, Wells, Somerset

again. This is called sgraffito, and will suggest new departures for you in other uses of enamel besides jewelry.

Another interesting experiment is to melt in the kiln on a piece of metal a child's coloured marble. This gives a more three-dimensional effect which is very jewel-like. These are a few methods you can try but they are merely an introduction to the many innovations and experiments you may conduct.

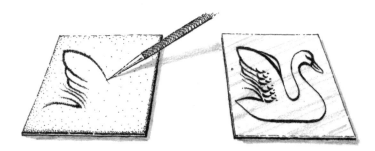

Fig. 48 Sgraffito. Scratch through the unfired enamel with a scriber to the glazed enamel beneath. By this method you may draw any shape you wish

8　Further techniques

Use of gold and silver foil

As your skill develops, you may wish to try the underlaying of gold and silver foils beneath transparent enamels. This is not only a very beautiful technique but it also enables the enamellist to achieve the effects of these precious metals without incurring the cost of the thicker gauges which would normally be enamelled upon. Gold and silver foils are usually obtained from your jeweller's supplier. They are available in booklets of twelve pages each approximately 4 ins square. Always be sure that you ask for enamelling foil. This is much thicker than the kind used by sign-writers etc. It is also almost pure metal, that is to say it contains virtually no other metals in its alloying.

Cut, prepare and counter-enamel a piece of copper or gilding metal 2 ins square. This counter-enamel should be fired at a low temperature sufficient only to fuse the enamel to the metal, not to glaze it. After cooling, clean the edges of the plate and prepare the other side. Enamel this with any of the transparent enamels. Then with a scriber prick any air bubbles that may have occurred on the transparent side and fill these, together with any other pits, with the same transparent enamel and fire to a smooth surface. Now draw a 2-inch square on a piece of paper and sandwich your first sheet of foil between this and another piece of paper behind, and cut out the square with shears or a razor blade. The paper prevents any unnecessary damage to the foil, and also enables you to mark out more complex shapes on it without having to mark the foil. The square of foil, called a *paillon,* must now be pierced with approximately eight hundred pinpricks, which is done by embedding a dozen or so needles into a cork and then dabbing the paillon firmly with this. The minute holes made in this way allow air trapped between the paillon and the enamel to escape and provide a slight key to the paillon to help it adhere to the enamel surface. With smaller paillons this may not be necessary but I would recommend it wherever the scale makes it practical.

Paint adhesive over the surface of the transparent enamel and carefully lay your paillon onto this. Dry it with blotting paper, absorbing as much moisture as possible before finally drying off at the entrance to the kiln. Check that the work is completely dry by holding it up to a dark background when it is still warm. If no moisture is present, it is safe to fire it. Be very careful not to melt

Fig. 49   Cloisonné box by Kenneth Bates, USA

Fig. 50   Circular cloisonné bowl by Kenneth Bates, USA

the paillon during the firing. Gold and silver melt at much lower temperatures than do gilding metal or copper. Remove the work when the foil appears to be a dull red. Immediately after it is removed from the kiln, and while it is still hot, use your stainless steel spatula to press down and burnish the paillon onto the enamel which will still be 'sticky'. This ensures that the paillon is firmly embedded in the enamel and that any trapped air is extruded. After this, refire the piece, paying particular attention again to the temperature of the foil.

Blues and greens look particularly attractive over silver, as do reds and yellows over gold. Apply a transparent enamel of the colour you have chosen, by the dusting method, as thinly as possible over the paillon. Pay particular attention to the edges, and then fire this. To fully exploit the qualities of gold and silver foils the enamels should be washed as thoroughly as possible and applied as thinly as possible. The use of foils is particularly effective in conjunction with the cloisonné method (page 61–3). Different 'compartments' may have gold and silver laid in them, exactly as described above. This gives an extra depth and lustre to the enamels.

Painted enamel

A further method which the reader should try at some time is painted enamel. For this, the enamels are ground to a specially fine frit and are much finer than the standard ground enamel. They are available from your supplier under the name 'painted enamels'. Some of these use water as a medium and others oil of lavender. Usually this is specified by the manufacturer.

This is a simple technique and a very small quantity of painted enamel will go a long way. Therefore, to start with, mix a tiny amount of painted enamel with the appropriate medium, water or oil of lavender, in a saucer. Using a good-quality number 2 sable hair watercolour brush, mix the enamel and medium further. Keeping the brush well charged with enamel, wipe away the surplus on the edge of the saucer to give a good point to the brush. Then, on a glazed enamel surface, carefully paint a linear design. This should not be too complex for your first attempt. You will find that, apart from being slower to apply, the enamel will apply very much as any other sort of heavy paint. Dry this at the

Fig. 51 *Owl* enamelled on steel tiles by Geoffrey Franklin, England. Combined use of dusting, stencil and spatula methods with opaque and transparent enamels, and also painted enamel

mouth of the kiln and fire until a slight glaze appears on the painted lines. Over-firing of painted enamels usually results in burning the enamel with the consequent loss of colour and glaze. Over-fired soft painted enamels tend to run or 'bleed'. I prefer to use painted enamels in either very delicate work or where a small amount of colour, or lines other than cloisonné lines, are needed. In almost all other cases it is better to apply ordinary enamel by the dusting or spatula methods, as it is difficult to obtain a larger flat area with painted enamels. Liquid gold and silver in small bottles are also available from your dealer. These require no added medium and are applied with the same sort of brush. Care should be taken, however, to shake the bottle well before use, as the gold and silver powders tend to sink through the medium to the bottom of the bottle. Fire as described for the painted enamels. Failures are also the same, except that a broken or fragmented line is usually the result of too much medium and insufficient gold or silver. Shaking the bottle is therefore one of those obvious things that is all too often neglected.

Trail enamelling

Trail enamelling is an exciting method of using enamel, particularly for small pieces such as jewelry (figs. 45 and 46). On a previously enamelled metal plate, place lumps of enamel of suitable colours. Fire these, and when the enamel is molten draw your sgraffito tool through the molten enamel. Care must be taken not to scrape through the original glazed ground coat to the metal beneath. If the sgraffito tool sticks to the enamel, give it a sharp tap with a file or pair of shears and it usually releases itself. Failing this, plunge the plate and the sgraffito tool into a bowl of tepid water. Cold water would shatter the enamel away from the metal plate. When scraping the sgraffito tool through the enamel make sure that the enamel is really molten and not just sticky. This gives a better effect and virtually eliminates the possibility of the tool sticking to the enamel. If you do not possess or want to buy cake lump enamel, as it is called, it can easily be made with the powdered enamel you already have. A good way of doing this for small quantities is to heap small 'pyramids' of enamel mixed with water on a metal plate. This plate should not be thoroughly cleaned as is usual for enamelling, but it should not be so dirty as to impart

impurities into the enamel. Dry these 'pyramids' thoroughly at the entrance to the kiln and put the plate into the kiln and glaze them at high temperature. As they glaze they will begin to sink and become more rounded. At this stage plunge the plate and enamel into cold water. The lumps of enamel will separate from the plate and, after drying, are ready to be used for trailing.

Opposite
Fig. 52   Chalice by Nicholas Wilson for Messrs. G. Harvey and Sons, England. The painted enamel stem won the second prize at the annual demonstrations of craftsmanship in the enamelling section organised by the Worshipful Company of Goldsmiths, 1963

Fig. 53   Trail enamelled dish by a schoolgirl aged 13, Wells, Somerset

Threads of enamel

Finally, a further technique of enamelling is the use of threads of enamel. These may be made by heating a lump of enamel until it is molten and then dipping the end of your sgraffito tool into it and drawing upwards. This produces thin threads of enamel which harden almost instantly. These threads may be cut or broken into appropriate lengths and laid on the surface of any enamelled plate and fired until sufficient fusion has taken place to adhere the two together permanently. When making the threads it is much easier if the heating of the cake lump can be done with a gas torch (see page 64). In the kiln, because of the large area of heat and the height of the kiln firing chamber, it is very difficult to insert the sgraffito tool and lift it upwards. Only short threads can usually be made in the kiln. The thread method is very effective, especially on pieces where you want to make use of the relief texture provided by the threads, for example on handles and the edges of ash trays.

With all these techniques there is no reason why you should not intermix the principles. These are a few methods I have found successful. I am quite sure they are not the only ways and I hope that as you gain experience you will find other ways of your own to increase your enjoyment of the craft.

Fig. 54   Mature student's work with rings set in the enamel

9 The use of enamels

When you have gained sufficient experience with the introductory exercises in this book you will no doubt seek further outlets for your newly found skills. Enamels may be exploited in very many ways, and you will find with continued experience and knowledge that the field in which they can be used will become wider.

Pieces of work for enamelling upon

Many readers, for a wide variety of reasons, will not wish to concern themselves with the making of forms to enamel upon. There is a wide range of manufactured pieces, specially prepared for the enamellist's requirements and made in suitable metals to the correct thickness. By using these, you can then concentrate your valuable time and efforts on the craft of enamelling.

The particular aspects of jewelry are dealt with in chapter 7, but more general pieces are available in copper and gilding metal. These include shallow bowls and trays of all shapes in small sizes; flat discs of all shapes up to 4 ins in diameter; match box covers, different shaped napkin rings and square tiles. One manufacturer supplies wooden pieces to mount your enamels upon. The real enthusiast will in time exhaust this range and in order to further this interest it is an idea to have pieces made up specially by local industry. This need not be as expensive as you imagine. Some firms in Great Britain will introduce copper or gilding metal into their existing production runs if they are approached in the right way. If there is a local firm with a spinning lathe they should be approached to see if they will produce a few bowls or suchlike for you. Things made by the spinning method are perhaps the cheapest, and it is certainly one of the quickest methods. Once the former has been made it is as well to have several bowls spun, as the cost of making the wooden former is the biggest single item.

Whatever you make or design to be made for enamelling upon, try to avoid sharp corners. As already stated, a gentle radius is always preferable. Avoid square boxes. Even with counter-enamelling the sides tend to distort in the kiln. It is usual to enamel flat panels and fix these to the sides of the box afterwards. Make curves gentle and avoid sharp pointed forms. Points tend to heat up quicker in the kiln than the other parts and the enamel burns out here earlier than is usual. Try not to join two pieces of metal together before enamelling. Soldering and riveting do not take kindly to enamel and contraction problems are likely to occur near these points. If solder is used, it should be silver enamelling solder which has a higher melting point than the enamel. If you remember that whatever you design or make you have to enamel afterwards, you will have a sound form. Try to analyse the enamelling problems while you are designing and you will save yourself a great deal of trouble afterwards.

Tools for making work to be enamelled

For those readers who do wish to make their own forms the following tools are an essential addition to those already specified:

First, a jeweller's saw (fig. 9, page 19) with a gross of fine blades. This is perhaps the most important tool. With this you will be able to cut out the many profiles and shapes that you will need. Remember that the teeth of the blade must face the handle, and put a reasonable amount of tension on the blade as you insert it in the saw frame, by leaning on the frame. The cut is always made on the down stroke. The blade does not cut on the return stroke.

Second, needle files, which should include square, three square, flat and rattail sections. With these you will be able to file small parts with accuracy (fig. 9).

Some short lengths of 1 in. and 3 in. diameter mild steel will be useful for tapping curves over with a mallet. Any further tools tend to be for specific projects and go beyond the enamelling context of this book. As a design principle the simplest pieces are invariably the most successful. This is a particularly relevant guide when you are making pieces for enamelling upon.

New uses for enamel

I will now deal with a few other items which offer a little more original use of enamel. Flat enamel panels may be mounted on table tops and trays of all sorts. It is not necessary to use new pieces or have them made specially for this. By cementing your enamelled tiles to old furniture, it may be given an exciting new look. Clearly the size of the tile used will be determined by the size of your kiln, but this should not deter you from undertaking, say, a coffee table. The multiple-tile table-top (fig. 56) has, in my own opinion, a certain quality which the more professional single sheet of metal, fired in a large kiln, does not have. The enamel top should

Fig. 55   Dusted enamel and cloisonné individual dining place mat. The circular plate is cemented to a suitable insulating material such as a cork or linoleum tile

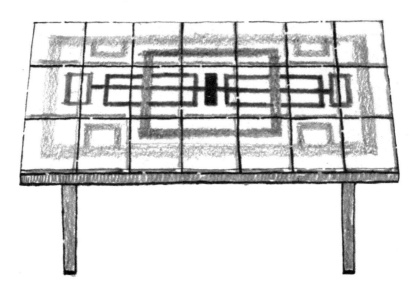

Fig. 56   An enamelled tiled table top. The rectilinear pattern enables you to apply the design accurately on the tiles

be protected by a sheet of glass to prevent sharp objects chipping the brittle surface of the enamel. The enamel tiles should be cemented to the table top with one of the modern impact (mastic) adhesives for gluing metal to wood.

The same method can be used for individual dining place mats, or any other flat surface which requires an easily cleaned and decorative finish. Depending upon the function of your flat surface, you may be able to dispense with the protective glass cover. Provided the enamelled surface is not going to receive harsh treatment this is quite possible. Old mirrors, photograph frames, etc., may be given a new look by cementing enamelled units directly to the edge of the glass. Cedar wood cigar boxes can be used by cutting plates of metal to fit the sides and lid and enamelling these. When cemented on to the box they can look most effective. Enamel is suitable for any place, such as bathroom or kitchen, where condensation or steam are a problem. As a decorative glazed splashback, enamel provides an excellent non-

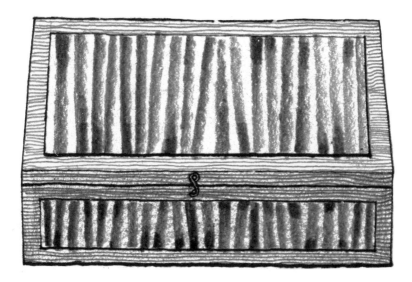

Fig. 57   A standard cigar box with enamelled panels applied to the lid and the sides

absorbent surface which will reflect heat and light and which is easily cleaned by simply wiping over with a damp cloth.

Enamelled panels are now used for external wall cladding. They have the advantage of insulating the building and require no maintenance. Normal rainfall keeps the surface clean and bright. It is therefore quite possible for you to enamel your own house name or number for mounting outside. It is better to cement the plate to a piece of wood or similar material which may be screwed to the wall. Always try to avoid bending or distorting the enamel as this can cause the enamel to shatter. Mosaics can also prove a good way of using enamel. For further information on this I can recommend *Making Mosaics* by John Berry (see 'Further reading', page 102). This covers the making of mosaics in a wide variety of media many of which could be substituted by enamels.

It is said that the painter or artist lies hidden in most of us. Enamel is obviously a wonderful medium for the artist, both amateur and professional, provided that the real nature of the material and its colour are appreciated. No comparison with paint of any sort can be made. Provided this is understood you may set about making abstract compositions with confidence. For the novice it

Fig. 58   An enamelled cloisonné house number plate. The plate is cemented to a wooden base board

is better to start with abstract subject matter to allow for the 'surprise' element in the firing. Later, when more control is achieved, there is no reason why you should not attempt more pictorial compositions. For those who find the drawing difficult, it is quite legitimate to cut out photographs from magazines for the profiles of your subjects, for use with the dusting technique.

It is a natural step from the picture or small composition to the mural. This may seem ambitious, but as the method is largely repetitive it is not necessarily so. Choose a small alcove first; the shower wall of a bathroom is an ideal location in view of the properties of enamels mentioned earlier. Calculate the size and number of tiles you require and have these cut mechanically by a guillotine (shear). This will ensure that your finished work will fit its intended situation, and that all the tiles are identical in size. Metal sections, such as angle aluminium, make good surrounds (frames) for your work, as do natural woods.

Other materials may also be used as supports or for enamel display purposes. Plastic is not unsympathetic but care must be taken with the choice of the plastic colour. As a general guide, black, white, grey or transparent plastics are the most suitable. Cork is very effective, as is sand and concrete and other coarse surfaces. Whatever you use, do not subordinate the enamel to the background.

I have illustrated a few pieces which are suitable for beginners to make and for those who do not have workshops or special skills, and which can be easily assembled and mounted. Finally there are the many manufactured metal pieces mentioned in chapters 7 and 9 for enamelling upon. With these and any other suggestions for enamelling, I recommend you use the brilliant colours with discretion and that you explore the effects of the transparents. No other colour medium will give such a quick range of permanent colour. Enamel, well applied, will last for over one thousand years and is available to anyone who cares to explore this stimulating and exciting art craft.

Fig. 59   Enamelled filtration plant mural, 5ft × 24ft in 42 sections. Edward Winter, USA

Over page
Fig. 60   *Enthroned Christ*. Enamelled steel mural, 12ft × 16ft, for St. Mary's Romanian Orthodox Church, Cleveland, Ohio. Edward Winter, USA

10   A brief history of enamelling

Enamelling was first known to the Greeks in the fifth century B.C. Sculptures were then inlaid with thin areas of enamelled gold. Later, in the fourth century, it is thought that the technique of cloisonné originated. It is difficult to assess the extent of enamelling at that or any other early time in history as so much work was lost, pillaged or destroyed. Enamels on precious metals were always subject to destruction as they could be melted down for the value of the plate. Enamels were being used in the British Isles by the third century A.D. Good examples of Irish and Kentish enamels remain, and fine examples of red and blue enamels are thought to have come from the West country. The Romans also left behind pieces of enamel work, examples of which may be seen in the British Museum. A method known as champlevé also evolved at about this time. This involved the pouring of molten opaque enamels into specially carved or hollowed areas of metal. This method is little used today except for badge and insignia work, but an example is shown in fig. 67, page 95.

Enamelling flourished in the Byzantine Empire during the ninth and eleventh centuries. Principally enamelled upon gold in the cloisonné manner, the work of the later half of the century is of incredible intricacy. The style of the work was dictated largely by the ecclesiastical nature of the subject matter. It is from Byzantium that the art is believed to have been introduced to Germany, by a Byzantine princess, Theophano, when she married the German King Otto II. She took with her Byzantine craftsmen who continued to work in their own native manner. German enamellists are known to have started combining the methods of cloisonné and champlevé in one piece of work, which hitherto had not been done. Examples of German work may be seen in the Cleveland Museum of Art.

Fig. 61   Enamelled bronze plaque from the river Thames, 2nd century AD. British Museum

The next major enamelling development belongs to France, and particularly to the town of Limoges which was to give its name to this exquisite technique. This method used a white groundcoat which was drawn upon in black oxide; over this would be laid numerous layers of transparent colours. At this time entire families followed trades, and the originator of the Limoges method was Leonard Pénicaud, sometimes known as Nardon. His brother Jean was also eminent in this field. Subsequently, Leonard Limousin and Pierre Reymond were great enamellists. Considerable attention was paid to the reverse side of a piece of work. Often it is quite difficult to decide which is the front, owing to the excellence of the counter-enamelling. Portraiture in enamelled miniatures at this time reached a high peak of achievement. These miniatures had often extremely good likenesses in spite of the difficult tech-

Fig. 62   Royal Gold Cup. French, c. 1380 AD. British Museum

nique of building up the tones in layers and firing these, and the exacting process of mixing the subtle skin tones. Important collections of miniatures may be seen in the Taft Museum in Cincinatti and the Victoria and Albert Museum in London.

In England a distinctive style developed in the workshops of Theodore Jansen at Battersea between the years 1750 and 1820. These were painted enamels on small boxes, bowls and other containers. Painted enamels are very finely ground and are mixed with a vegetable oil to act as the medium. The Battersea enamels often depicted elegant people dressed in the clothes of the times, set in misty woodland scenes. Birds and flowers were delicately painted on white backgrounds, nearly always after the style of the painter Watteau.

Russian enamelling is believed to have been introduced also

Fig. 63 *The flight into Egypt.* Enamelled plaque by Jean II Pénicaud, Limoges c. 1540. Victoria and Albert Museum

Fig. 64   King John Cup, English 13th Century. Kings Lynn Corporation

Fig. 65   An incredible gold and enamel chalice-cup by Miss H. M. Ibbotson, England, 1927. The plique-à-jour method is usually confined to flat pieces. Here, by tremendous skill and control, the transparent enamels are retained in thin gold wires completely in the third dimension

from Byzantium; it certainly carries the stylistic influences. Plique-à-jour is a Russian development. This very beautiful method is achieved by firing enamel in between fretted areas of metal but without a metal background. Mica paper is used to support the enamel during application and firings, and is later peeled away. The finished effect is jewel-like and may be compared with a miniature stained glass window. The technique is little used today as its effect depends upon light shining through it from behind; its main use now is for altar crosses and suchlike.

Chinese enamel is mostly concerned with cloisonné enamelling on brass. The characteristic oriental style never ceases to amaze with its great detail and strong rhythmic lines. The vast quantity of this work probably accounts for its very low market value, although much of it is not perhaps as well made as its general appearance would lead one to think. Often the scroll ends of the cloisons are not fully enamelled and small pits and holes may be found on close inspection. Much Japanese cloisonné work was not fire-glazed after stoning back. The final effect is a pleasant egg-shell finish but slightly porcelain in character.

Fig. 66   Ming cloisonné jar, c. 15th century, Victoria and Albert Museum

Fig. 67   Cigarette box designed by Alex Styles and made and enamelled by Messrs. Padgett and Braham, England, 1965. A strong, simple geometric pattern beautifully executed in champlevé enamel

Fig. 68   An exquisite dragonfly brooch by René Lalique. Made in the Art Nouveau
period in cloisonné enamel with added chrysoprase

From the middle of the nineteenth century a revival of enamelling started in England. Largely as a reaction to the advance of the industrial revolution and its many implications, figures such as Harold Stabler, H. H. Cunnynghame and Alexander Fisher emerged as staunch defendants of the traditional methods of handicraft enamelling. It is largely due to their efforts that enamelling in Europe today is as popular as it is.

No chapter of this sort can be written without paying tribute to the enormous contribution made in modern times by the enamellists of the United States. It is no exaggeration to say that these men and women took the enamelling world by the collar and shook it, particularly in the use of enamels. One could write a long catalogue of the names of people involved in this minor renaissance; but inevitably someone would be neglected and this would be wrong. Within the abundance of work produced in America came the development of new techniques which are now standard studio practice. Enamels for aluminium are being developed, and I have no doubt that other significant artists and their methods have yet to emerge.

Notes on enamel behaviour

The following notes are generalizations on a few colours and are not intended as firm guides to all enamel behaviour. These will vary from one manufacturer to another and from one enamellist to another. Each individual will work the enamel differently, so no firm results can be predicted. The terms 'soft' and 'hard' indicate a low or high melting point enamel respectively. Generally speaking, hard enamels tend to be acid-resistant and soft enamels are not.

Blues The most popular and fault-free of all the enamel colours. There is a wide range of colours in both the opaques and transparents, all of which tend to resist cracking and to be free of pit holes. The hard types are particularly long lasting and are acid-resistant. When applying transparent blues on copper, they should be applied a little thicker than usual to achieve their true hues.

Browns A good range of reliable colours.

Greens Transparent greens, like the blues, should be applied a little thicker than usual when fired upon copper, to achieve their true hues. Opaque greens have a tendency to have small pit holes in the glazed surface.

Greys These are inclined to have a coarse surface caused by pit holes in the glaze. These holes may be filled and refired to obtain a smoother surface.

Purples A limited range only of opaques. There is a wider range of transparents. When fired upon silver there is a slight tendency towards green in the final colour.

Reds There is a reasonable range of reds available but it is the most difficult of all the enamels to fire. All reds are soft and as such tend to burn out quickly when fired. The washing of the enamel and the preparation of the metal are most important and fire scale should be guarded against. Transparents will require two or three coats over a ground coat of clear, transparent flux.

Whites A wide variety of hardness, designed for various purposes. When soft white is used on copper it tends to leave green and/or bluish tints on the finished surface. When soft white is used for thin lines it will bleed if over-fired.

Yellows As a group, these tend to hardness with a critical glazing temperature. If over-fired they are inclined to burn out. The transparent yellows are slightly cloudy when fired and are not entirely transparent.

General Many other colours have their variations and subtleties. The so-called faults of the colours mentioned above and of other colours may be adopted as virtues to produce certain effects in your work. Test firing as mentioned in chapter 5 is the key to this exploitation and is again an essentially individual process.

Problems and how to avoid them

Fault	Reason
Thin sparse appearance of enamel.	Enamel applied too thinly and probably unevenly. This is very obvious on transparent enamels.
Slightly porous-speckled appearance of the enamel.	Either the enamel has been applied too thinly, or it has been under-fired, or possibly both.
Small round volcano — like holes in surface of glazed enamels.	Enamel was fired slightly damp.
Enamel lifting when entering the kiln.	Enamel was fired damp.
Edges of enamel curling and rolling back.	Edges of metal not clean and the enamel over-fired.
Areas of enamel flaking away after firing.	Metal not clean and the enamel cooled too quickly after firing.
Cracks appearing.	Enamel cooled too quickly. Also possibly due to inadequate counterenamelling, thus allowing the metal to distort.
Black specks appearing on fired enamel.	Fire scale allowed to settle on fired and prepared enamel surface.
Positive change of colour after firing sequence.	Enamel has been over-fired consistently. This destroys the nature of the colouring oxides in the enamel.
Transparent enamels appearing slightly cloudy and not fully transparent.	Enamel not washed sufficiently and possibly applied unevenly.
Enamels lifting off in layers when cooling.	Either (a) A low-temperature firing enamel has been fired over a high-temperature one. Although the final enamel may have glazed, the underneath enamel has not, and no fusion has taken place.
	or (b) The enamel was cooled too quickly after firing.
	or (c) The surface of the enamel was not cleaned before the second layer was applied.
	or (d) Any combination of a, b or c.

List of suppliers

Suppliers
Enamels
B. K. Drakenfeld & Co., 45 Park Place, New York 7
Ferro Enamel Corporation, 4150 East 56th Street, Cleveland, Ohio
Metal Crafts Supply Company, Providence, Rhode Island
The O. Hommel Company, 209–213 Fourth Avenue, Pittsburgh 30, Pennsylvania
Thomas C. Thompson, 1359 Old Deerfield Road, Highland Park, Illinois 60035
American Art Clay Co. Inc., 4717 West 16th Street, Indianapolis, Indiana 46222
Allcraft Tool and Supply Co. Inc., 15 West 45th Street, New York, NY 10036

Kilns
Anchor Tool and Supply Company, 12 John Street, New York
Broadhead Garrett Company, Cleveland, Ohio
Electric Hotpack Company, Coltman Avenue, and at Melrose Street, Philadelphia, Pa.
Hevi Duty Electric Company, Milwaukee I, Wisconsin
American Art Clay Co. Inc. (as above)

Jeweller's supplies
General Findings Inc., Leach & Gardener Buildings, Attleboro, Massachusets 02703
B. A. Ballon & Co. Inc., Peck and Friendship Streets, Providence, Rhode Island 02903
Magic Novelty Co. Inc., 95 Morton Street, New York, NY 10014
T. B. Hagstoz & Son, 709 Sansom Street, Philadelphia, Pa. 19106
Thomas C. Thompson (as above)
Allcraft Tool and Supply Co. Inc. (as above)
Orange Brothers, 40 John Street, New York, NY 10038
Sweet Manufacturing Co., West Mansfield, Massachusets 02083
Novel Products Co., 42–61 24th Street, Long Island City, NY 11101
City Novelty Co., 450 West 31st Street, New York, NY 10001
E. A. Adams & Son Inc., 545 Pawtucket Avenue, Pawtucket, Rhode Island 02860

For further reading

Enamelling: Principles and Practice by Kenneth F. Bates; Constable, London, and World Publishing Co., Cleveland and New York, 1951

Making Mosaics by John Berry; Studio Vista, London and Watson-Guptill Publications, New York, 1966

The Preparation of Precious and other Metalwork for Enamelling by De Koningh; Norman W. Henley Publishing Company, New York, c. 1930

Simple Jewellery by R. W. Stevens; Studio Vista, London and Watson-Guptill Publications, New York, 1966

Index